MIX
Papier aus verantwortungsvollen Quellen
Paper from responsible sources
FSC® C105338

Adeolu Aderoju

TECHNOLOGICAL ADVANCEMENT IN THE OIL AND GAS INDUSTRY

A CONSIDERATION OF THE NODAL SEISMIC SYSTEM

Anchor Compact

Aderoju, Adeolu: TECHNOLOGICAL ADVANCEMENT IN THE OIL AND GAS INDUSTRY:
A CONSIDERATION OF THE NODAL SEISMIC SYSTEM, Hamburg, Anchor Academic
Publishing 2015

Buch-ISBN: 978-3-95489-394-2
PDF-eBook-ISBN: 978-3-95489-372-0
Druck/Herstellung: Anchor Academic Publishing, Hamburg, 2015

Bibliografische Information der Deutschen Nationalbibliothek:
Die Deutsche Nationalbibliothek verzeichnet diese Publikation in der Deutschen
Nationalbibliografie; detaillierte bibliografische Daten sind im Internet über
http://dnb.d-nb.de abrufbar

Bibliographical Information of the German National Library:
The German National Library lists this publication in the German National Bibliography.
Detailed bibliographic data can be found at: http://dnb.d-nb.de

All rights reserved. This publication may not be reproduced, stored in a retrieval system
or transmitted, in any form or by any means, electronic, mechanical, photocopying,
recording or otherwise, without the prior permission of the publishers.

Das Werk einschließlich aller seiner Teile ist urheberrechtlich geschützt. Jede Verwertung
außerhalb der Grenzen des Urheberrechtsgesetzes ist ohne Zustimmung des Verlages
unzulässig und strafbar. Dies gilt insbesondere für Vervielfältigungen, Übersetzungen,
Mikroverfilmungen und die Einspeicherung und Bearbeitung in elektronischen Systemen.

Die Wiedergabe von Gebrauchsnamen, Handelsnamen, Warenbezeichnungen usw. in
diesem Werk berechtigt auch ohne besondere Kennzeichnung nicht zu der Annahme,
dass solche Namen im Sinne der Warenzeichen- und Markenschutz-Gesetzgebung als frei
zu betrachten wären und daher von jedermann benutzt werden dürften.

Die Informationen in diesem Werk wurden mit Sorgfalt erarbeitet. Dennoch können
Fehler nicht vollständig ausgeschlossen werden und die Diplomica Verlag GmbH, die
Autoren oder Übersetzer übernehmen keine juristische Verantwortung oder irgendeine
Haftung für evtl. verbliebene fehlerhafte Angaben und deren Folgen.

Alle Rechte vorbehalten

© Anchor Academic Publishing, ein Imprint der Diplomica® Verlag GmbH
http://www.diplom.de, Hamburg 2015
Printed in Germany

ABSTRACT

Technology has proved its credibility by helping us to combat some of the most important challenges in decades. On a high interest, our yesterday's concerns are now our jubilations today. Also, various innovations that technology is offering our industry today has shown to us clearly that the industry cannot afford to shuttle its today's cares till tomorrow. It is even more interesting that today, creative minds in the industry are already gazing into the future in other to tackle our tomorrow's challenges right from now. All these validate a simple fact that: "the oil and Gas industry is a technology based and innovation driven" (David, 2011).

The phase changes and its adaptation is so rapid that if professionals fails to yield to it or feels reluctant to its tingle, such would be dismayed in waiting and may find it stiff in catching up with the transiting train. The thrust of the drive witnessed by the industry in the recent decade is intense. This is wholly responsible to the world's high demand for our commodity (Oil and Gas), our daily venture into the ultra-deepwater exploration, unconventional resource exploration systems which is always beckoning on new and strong techniques. All of these are enough to charge professionals, to be awake to the demands of their fields by yielding their thoughts to the present breakthrough and preparing to face the next decade's challenges

One important breakthrough that technology has offered the seismic data acquisition field of recent is the "Nodal Seismic System". The success is currently attracting a great deal of key players from every end of the industry and has kept discussions on over time. The advancement is also known as Cableless, Wireless or Nodal seismic acquisition system, as it may be. It is an improvement over the conventional cabled seismic acquisition system. An overview of this advancement In relation to the challenges it solved has been looked into in his article.

Table of content

ABSTRACT 5
Table of content 6
INTRODUCTION. 7
CABLE FREE SEISMICS 9
General Statement 9
Cable-free land seismic Data acquisition 9
Problem Statement 10
THE MECHANISM DEFINED 13
CASE STUDY 24
CONCLUSION 28
References/Bibliography 29

INTRODUCTION.

Technology has proved its credibility by helping us to combat some of the most important challenges in decades. On a high interest, our yesterday's concerns are now our jubilations today. Also, various innovations that technology is offering our industry today has shown to us clearly that the industry cannot afford to shuttle its today's cares till tomorrow. It is even more interesting that today, creative minds in the industry are already gazing into the future in other to tackle our tomorrow's challenges right from now. All these validate a simple fact that: "the oil and Gas industry is a technology based and innovation driven" (David, 2011).

The phase changes and its adaptation is so rapid that if professionals fails to yield to it or feels reluctant to its tingle, such would be dismayed in waiting and may find it stiff in catching up with the transiting train. The thrust of the drive witnessed by the industry in the recent decade is intense. This is wholly responsible to the world's high demand for our commodity (Oil and Gas), our daily venture into the ultra-deepwater exploration, unconventional resource exploration systems which is always beckoning on new and strong techniques. All of these are enough to charge young professionals, to be awake to the demands of their fields by yielding their thoughts to the present breakthrough and preparing to face the next decade's challenges

One important breakthrough that technology has offered the seismic data acquisition field of recent is the "Nodal Seismic System". The success is currently attracting a great deal of key players from every end of the industry and has kept discussions on over time. The advancement is also known as Cableless, Wireless or Nodal seismic acquisition system, as it may be. It is an improvement over the conventional cabled seismic acquisition system.

At first, the urge for a new technology as a replacement for cables was bare because the use of cable for seismic exploration was a natural, common-sense choice. It was what was available to meet the acquisition needs of the time. Even then, there was no need for a new technology because most explorations are done in only accessible locales, but now that "all the easy oil has been found", there was a great call for a new technology and the need has been met through the Cableless seismic system. An overview of this advancement In relation to the challenges it solved has been looked into in his article.

Many are the other success that has been recorded through technological advancements (although it would not be considered as such); improvements in Geophysics have led to the clear understanding of reservoirs. More often than before, Petrophysicists analyze reservoirs more confidently than before. Also, technological improvements have helped to have a much better picture of subsalt prospects in important oil reserves of the world. It was formerly believed that targets below salts are subsalt and are difficult to image while targets below limestone are lime.

Other Technological enablements are;

> The realtime reservoir modeling (4D Seismics, Intelligent Completion Downhole sensors), Well productivity (new well geometries) and many more.

Presently, the industry is being faced with some challenges which are yet to be efficiently catered for by these advancements. Productions and understanding of shale gas, shale oil and other unconventional resource systems still need some advancement. One can afford to put the mind at rest because more companies and individuals are delving more into researches on these areas of less understanding.

The question that could begin to bother the mind is that: "where would the fast transiting train of technological advancements take us to in the next decades?" a definite answer to this question might be hard but, undoubtedly, the train is taking us to heights. It is great to watch but more great to join the change. As for me, I care to be a part of the movement. What about you?

CABLE FREE SEISMICS

From the introduction it has been said that, this seminar work would concentrate on the cable-free seismic systems

About Wireless Seismic
Wireless Seismic was formed in 2006 to develop and introduce a revolutionary seismic data acquisition system to the exploration and production industry, capitalizing on emerging technologies in the seismic, wireless and mesh-network industries. Its financial backers include Chesapeake Energy Corporation, one of the largest producers of natural gas and one of the largest users of seismic data in the United States, and Energy Ventures, a Norwegian-based venture capital firm with focus on investments in the upstream oil and gas market.

General Statement

We all know how cellular wireless telephones have spread around the world. "Cell" phones are in every nook and cranny of the earth and are used by people of all ages, nationalities, and professions. This same cellular wireless technology has now entered the onshore seismic data-acquisition world. Just as a distant friend using a cell phone can cause a system of radio-tower relays to reach your cell phone and leave a message or transmit a graphic image, a small cellular wireless unit attached to a geophone can transmit the data recorded by that geophone through a system of radio antennae to a central data-storage unit.

Cable-free land seismic Data acquisition

Cables have always been seen as a necessary evil when it comes to seismic surveys, but the good news is that the burdens created by cables may soon be a thing of the past (Dennis, 2011).
Starting with the very first crew, all land seismic acquisition projects have had one characteristic in common: Cables, cables and more cables. The land seismic data acquisition industry is known around the world for requiring massive amounts of cable.

The terrain over which these cables must cross is often treacherous; the weather, sometimes harsh; and the equipment, always heavy and cumbersome. Cables have always been a burden, not only to acquisition crews, but also to the land being surveyed and, ultimately, to the individuals who use the data acquired.

Earlier on, large moving mass, coiled seismic sensors, often called jugs, were wired together and connected to a magnetometer so that physical motions in the earth could be transformed into squiggles on graph paper. Today's sensors, from highly sophisticated velocity sensors to cutting-edge accelerometers, have dramatically improved. Likewise, acquisition output has evolved from those raw analogue squiggles to digital representations on magnetic and optic media. Unfortunately, despite these significant technological advances, one thing did not change. Until very recently, the vast majority of seismic data acquisition systems still required tons of unwieldy and troublesome cables.

Today, new options are being introduced that can partially or completely eliminate cables from the seismic data acquisition experience. Geophysical equipment manufacturers are developing acquisition systems that make cables obsolete, and not just the telemetry cables that carry data from the various remote units to a central location, but also the remote-unit power cables, as well as the analogue cables that transmit signals from the sensors to the remote units. Some systems even eliminate the need for a string of geophones from the equation. These new cable-free systems are commonly referred to as nodal data acquisition systems. Nodes are generating a great deal of attention from both contractors and oil companies worldwide because they represent one of the most important technological advances in the history of the industry.

Problem Statement

To quantify the amount of equipment required of a conventional cabled land system (and therefore what could potentially be eliminated by a cable-free system), let's analyse a typical 4,000-channel seismic acquisition system with receiver-station spacing of 30 meters (plus 10 per cent): Such a system would require 132,000 m. (132 Km) of in-line cable.

This cable would house telemetry twisted-pair wires and, in most instances, power wires. The in-line cable alone would weigh in at 6,000 Kg. Add in the geophone strings (33 m, each with six geophones) for another 132 Km of cable, weighing a bit over 15,000 Kg. Power for the line would typically require some 84 power stations at 58 Kg each for another 4,872 Kg. For our hypothetical 4,000-channel 2D line, we would require some 264 Km of cable and weigh in at around 25,872 Kg, or slightly less than 6.5 Kg, per channel. Add to this the fact that every receiver station would require three connections, two for telemetry/power and one for the sensor, and that each power station would need at least one connection, and the total number of connectors reaches over 12,000. That is a lot of equipment to protect and maintain, and this scenario does not even consider any cross-line requirements for 3D operations.

Even so, weight is not the only bothersome issue caused by seismic cables. Another is pure mathematics. There is either too much cable or not enough, and they must always be cut to specific lengths, or takeout intervals. If there is not enough cable, the project must be redesigned to accommodate the limits of the equipment at hand. If there is too much, extra equipment must be deployed to the field, so that the excess cable can be piled or coiled between receiver stations, which creates an environmental and operational burden, and can lead to seismic noise problems due to leakage.

To further complicate matters, cables are easily damaged by natural and cultural causes, and connectors wear out. When that happens, the field crew must often cease production to repair or replace the defective cable(s) and connector(s). Troubleshooting the spread becomes an ongoing challenge, which means that even when a seismic crew is not acquiring data, they are still costing money. Lost production is lost time, and lost time is lost money (See Figures 1, 2, and 3).

Fig 1: Traditional Cabled System being staged prior to deployment (culled from Oil and Gas Technology-Dennis F.)

Fig 2: Showing the cables repair shop in production

Fig 3: Showing the deployment of cabled acquisition system. (culled from Oil andGas Technology-Dennis F.)

Again, the good news is that the burdens created by cables may soon be a thing of the past. The modern seismic contractor finally has options that simply were not available even just a few years ago. Currently, there are a number of 'minimal cable' systems commercially available. These systems do not require cables to interconnect the individual seismic data acquisition units to a central recorder. As such, they are somewhat similar to radio telemetry systems of the past; in fact, some use radios to transmit quality control data to a central station or field data collection reader and, therefore, may require a radio license. While these 'minimal cable' systems do indeed eliminate the need for any telemetry cable, they still require cables to interconnect the individual pieces of equipment, such as a battery and sensor(s) to a remote unit. These connection cables, however short, are still prone to the same hazards as the conventional seismic cables of the past.

THE MECHANISM DEFINED

The scope of this seminar3 would broadly divide the cable-free seismic data acquisition systems into two types based on their mechanisms, namely:

1. Wireless Geophone Network
2. Autonomous Nodal System.

The Mechanism defined
A system that acquires seismic data using cellular wireless technology is similar to a cellular telephone system in a large city. Inside the hypothetical city limits, several radio towers create overlapping reception/broadcast areas that combine to cover the city. Through a connection of radio towers, a cellphone user at A can talk to B, or transmit digital information to, a second cellphone user at B. (Figure 4a), The diagram implies that A and B exchange information via pass-along communication links 1, 2, 3, and 4, which span many miles.

In wireless seismic data acquisition, a geophone is connected directly to a small, wireless, remote acquisition unit (RAU) that functions essentially the same as a common cell phone (Figure 5). The RAU has an accurate internal clock that is synchronized with the internal clocks in all other RAUs across the seismic spread.

Each RAU also has an internal GPS receiver that adds precise earth coordinates to all data acquired by its assigned geophone. The seismic signal from the geophone is digitized by the RAU and then stored in flash memory – the same type of
Memory used in cell phones functioning as cameras that acquire, transmit, and receive photographs. Wireless cellular seismic systems made by current manufacturers differ in how they handle the data received from geophones.
In some systems, each RAU transmits its data to a central data storage unit via a system of overlapping radio-antennae patterns. In Figure 4b, the data transmission from geophone station C to data-storage unit D occurs via pass-along protocols between radio antennae a, b, c, and d.

In other systems (Autonomous Nodal System), data stay in the RAU and are downloaded to a data-storage unit at appropriate time intervals. In one option, each RAU is physically transported to a local data-storage device and then returned to its assigned geophone station. In yet other systems, a technician visits each RAU at selected times with a PC and uses a data wand to dump data from the RAU memory into the PC.

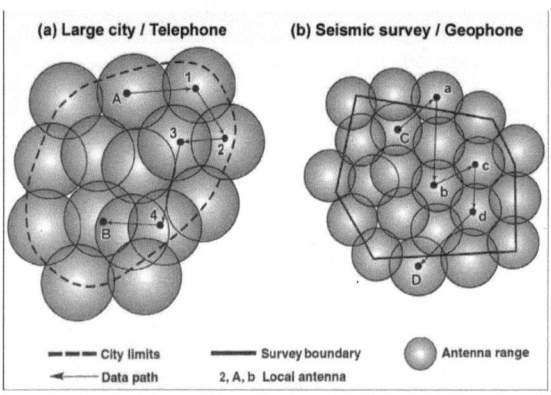

Figure 4: (a) Comparison between a cellular wireless telephone system spanning a city (b) and the same technology used to acquire seismic data.

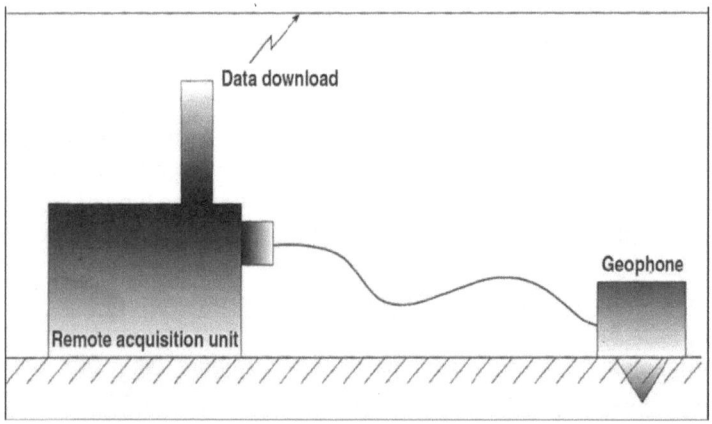

Figure 5: The principle of seismic data acquisition using cellular wireless technology. The remote acquisition unit (RAU) is, in essence, a cellular telephone with a huge memory that is built to

withstand harsh weather and rough treatment. The unit also has an accurate internal clock and a precise GPS receiver.

An RAU connects directly to a geophone string. The geophone output signal is digitized by the RAU and then is downloaded via radio links to a central data unit, or it is retrieved by a visiting technician, who downloads the data at the geophone station, or the RAU is transported to a data-download station and then returned to the geophone station after emptying its data.

1. Wireless Geophone Network

The logistic and weight costs that are caused by cables for high density receivers can be in principle removed by deploying geophones equipped with wireless trans-receivers to form a Wireless Geophone Network (WGN). A WGN might consist of units (receivers or wireless geophones, WG) that are in charge of sending their own recording data or forwarding (relaying) the data of other WGs.

Each WG is battery-powered and is typically equipped with a radio transceiver (for radio transmission and reception), small microcontroller and storage unit to handle processing, digitalization and buffering of seismic data. As for cable-based system, real-time data delivery needs to be guaranteed so that geophone digital data readings could be conveyed to storage unit with stringent delay constraints. Moreover, the control/storage unit provides with the necessary functions of timing and monitoring of each (wireless) geophone.

Wireless Communication technology

The large field extension and receiver density require the WGN to exploit sophisticated radio transmission technologies to efficiently handle either short-range transmissions (e.g., for receiver-to-receiver short-distance communication within one group interval) and long-range transmissions (for seismic data delivery to storage unit and geophone remote monitoring) that must cover distances of several kilometers.

From a communication perspective, a Wireless Geophone Network can be based on a mixture of network technologies that are working in cooperation: groups of WGs (e.g., within a group line) are forming independent wireless sensor networks (WSN, see Figure 4) that are simultaneously operating (e.g., sensing and transmitting using different frequencies/channels). At the same time, each network of sensors is interconnected by long-range wireless links to form what is usually referenced as scalable wireless metropolitan area network (WMAN).

WMAN network must support long-range links to collect the data traffic generated by each WSN for propagation toward the central control/storage unit.

Power Consumption

A limiting constraint in WGN is that sensing units must use very little energy. Differently from cable-based systems, a wireless geophone must be equipped with rechargeable batteries. Batteries might become the heaviest element of cable-free hardware as deployed wireless geophones may need to be left unattended for days. A fundamental goal of network design is thus to provide protocols and technologies that help in preserving energy consumption to maximize time between recharging. This particular issue is a key point that needs to be addressed at all layers of communication systems, from hardware to network layer and application. Power consumption of a WG is primarily ruled by transmitting/receiving circuitry, while energy used during acquisition and digital conversion can be regarded as negligible.

A typical example of this is the RT 1000 System developed by: Wireless Seismics Inc.

Overview of RT 1000 System

(A) System components
    a. Wireless Remote Acquisition
        Units (Picture from www.wirelessseismics.com)

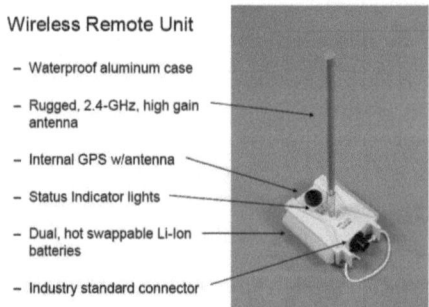

- Wireless Remote Unit
  - Waterproof aluminum case
  - Rugged, 2.4-GHz, high gain antenna
  - Internal GPS w/antenna
  - Status Indicator lights
  - Dual, hot swappable Li-Ion batteries
  - Industry standard connector

Fig 6: Wireless Remote Unit (Picture from www.wirelessseismics.com)

    b. Wireless data backhaul

Fig. 7: The wireless data backhaul. This serves as the local control for all nodes in the node's vicinity. First hand node data information is received at the data backhaul. (Picture from www.wirelessseismics.com)

c. Central recording system

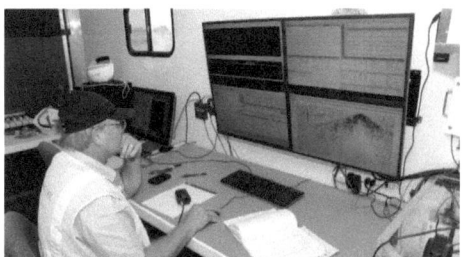

Fig 8: Central recording system- data information from the data backhaul is received at the CRS and later subjected to further processing. (Picture from www.wirelessseismics.com)

(B) Real Time Delivery
  a. Real time noise monitor
  b. Continuous QC
  c. No physical data collection or transcription
  d. No loss of data via theft or loss of box
  e. Seismic data recorded in Real Time

A (1). Wireless Remote Unit
- Waterproof aluminum case
- Rugged, 2.4-GHz, high gain
- Antenna
- Internal GPS w/antenna
- Status Indicator lights
- Dual, hot swappable Li-Ion batteries
- Industry standard connector

Deployment
- Deployment is easy and quick to learn
- Deployment requires minimal skill

Fig 9: Demonstrating the deployment of the RT 1000 system (Picture from www.wirelessseismics.com)

Batteries & Charging Station

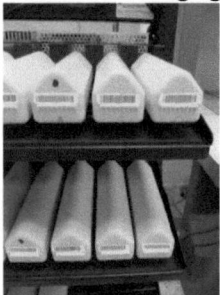

Fig 10a: Battery and Charging rack (Picture from www.wirelessseismics.com)

Fig 10b: Battery and Charging rack (Picture from www.wirelessseismics.com)

Smart Batteries and a dumb charger

• Li-Ion battery monitors own charge level, tells when it is charged, when in use, tells central how much power remains

Wireless Data Backhaul
– Commercial 5.8 GHz radio
– Mast is man-deployable in 10 minutes
– Operates on standard batteries
– Each BSU supports many WRU'

2. The Autonomous Nodal System

The Autonomous Nodal system is another type of cableless system which differ in its mechanism from the afore mentioned type. It is a more streamlined option. The autonomous nodes are completely cable free, with all of the essential elements to acquire seismic data contained within the nodes. Each lightweight, self-contained node encapsulates the sensor, batteries, control circuitry, A/D converter, filters, memory and a highly accurate clock to maintain timing. This approach allows the seismic contractor to offer clients ultimate flexibility in layout design, while simultaneously reducing their exposure to the many hazards posed by deploying a conventional system with its onerous bundles of cables.

This autonomous system presents a set of definite benefits and features (Louise M, 2011).

- Cable-free maneuverability
- Self contained sensing and recording
- Autonomous operation
- Exceptional reliability
- Continuous recording

A typical autonomous nodal system is comprised of these key components
- Node
- Source coordinator with safe. (safe and forward environment)
- Handheld Terminal (HHT)
- Data Recording Station (DRS)

1. Node – the node records filter sand stores a single point of seismic reflection point over an extended time period without cables of any kind.

Fig. 11: A typical Node. (Picture from www.fairfieldnadal.com)

2. Source Coordinator with safe – the source coordinator's SAFE device provides a GPS timestamp for each sources time break with accuracy within +/-1 microsecond of UTC. It is used to partition the nodes continuously recorded seismic data into field records.

3. Handheld Terminal (HHT) – Is a GPS positioning device using propriety Software to position, initiate, deploy and test the Node when laying out the receiver spread.

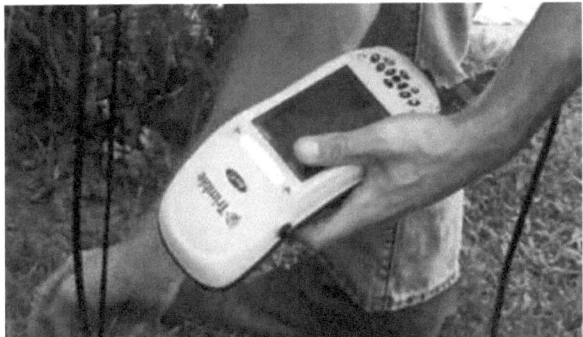

Fig. 11: A typical Handheld Terminal (Picture from www.fairfieldnadal.com)

4. Data Recording Station (DRS) – the DRS includes the data harvester section is where the Node and HHT's are checked in and out. Once deployed, each nodes precise location and time stamp are logged into the system to ensure accurate data

Fig. 12: A typical data collection rack (Picture from www.fairfieldnadal.com)

Fig. 13: A data recording station. (Picture from www.fairfieldnadal.com)

A typical example of this system type is the ZLand^R produced by the FairField Nodal Industries.

Overview of ZLand systems.

Fig. 11: A typical Node. (Picture from www.fairfieldnadal.com)
ZLand
It is lightweight, compact, cable free system, for rapid development and high production. It is especially well suited for use over rugged terrain and in urban areas, as well as on farmlands during growing and harvesting seasons.

ZLand is a nodal system capable of acquiring 288 hours or 12 days of continous seismic data. Cycling individual nodes in and out of a sleep state at predetermined times can extend deployment time significantly.

Node specifications

Mechanical
- Weight: 4.8 lb (2.2 Kg)
- Height: 6 in (15.2 cm)
- Diameter: 5 in (12.7 cm)
- Detachable 5-in spike (12.7 cm)

Electrical
- Timing accuracy
- GPS disciplined to +/- 100 microseconds UTC

Power
- Lithium-ion batteries
- Longevity: 288 hours
- Recharge time: 5 hours - worst case
- Flash memory Capacity: 2 GB
- Status LED indicating- Power ON, acquisition error

Data recording station specifications
Data harvester
- Mechanical
- Weight:
    - Empty ~ 250 lb (113 Kg)
    - Full ~ 460 lb (209 Kg)
- Capacity: 48 nodes
- Height: 90 in (229 cm)
- Width: 48 in (121.9 cm)
- Depth: 21 in (53 cm)

An individual data collection and charging rack (DCCR) can process up to 48 nodes. With a five-hour turn around, each DCCR can process up to 192 nodes per day. Multiple DCCRs can be joined together into a network to increase harvesting capacity.

Data sorter
- Mechanical
- Equipment rack
- Height: 54.5 in (138 cm)
- Width: 24 in (61 cm)
- Depth: 29 in (74 cm)

CASE STUDY

Due to the tenderness of this technology, much of its use has not been recorded in Nigeria. Although in 2011, Shell Exploration and Production Company did a 3-D Seismic survey using the technology to map an existing oil field where they formerly had challenges in mapping due to treacherousness of the terrain. Information about this survey could not be retrieved because the Project is proprietary.

However, many are the outstanding instances that can be referenced, where the new technology has been used and a good credit was given to the technology.

Frigid Siberia, Russia.

Just in 2011, (Louise. et. al, 2011), the Fairfield Nodal (mentioned above) did a field demo survey using the ZLand at the Frigid Region of Siberia (Russia). The survey has the distinction of having absolutely no cables of any kind. However, for comparison studies, the nodal implementation was performed alongside a 428 cable system, which is the convectional system.

A total of 202 nodes continuously records seismic data for four- plus days in snow covered forests where temperature dropped as low as -22^0C. The nodes were spaced 50 m apart on a 2D line, alongside a 428 cable layout. Trees reached as high as 17 m and the snow was as much as 80 cm deep. Even though the nodes were buried at about 10-50 cm in the snow, the nodes all functioned perfectly. It was recorded that the deployment was rapid, supervisor coupling in the snow and high quality data was collected. (Louise S. et. al 2011).

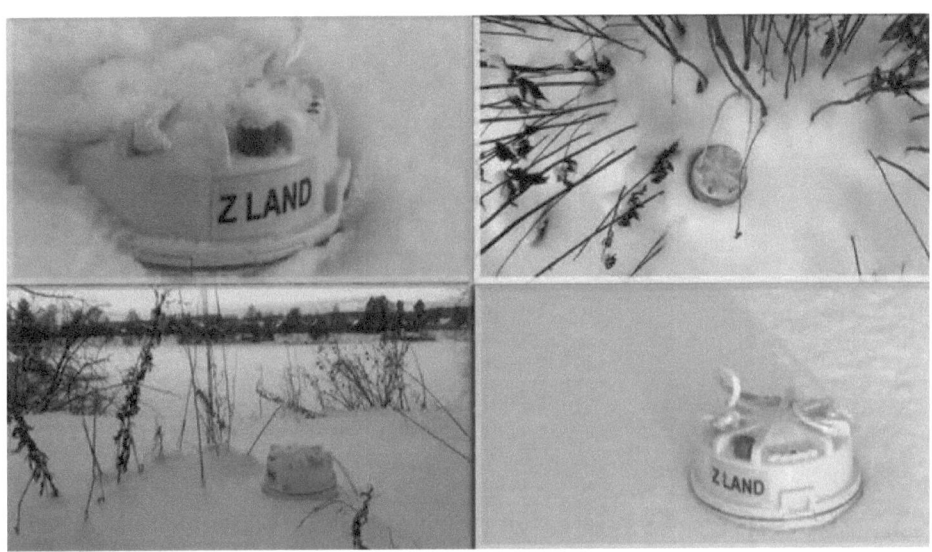

Fig. 12: ZLand being used in the snow, Frigid Siberia, Russia. (Picture from: www.fairfieldnodal.com

Fig. 13: ZLand being used in the snow, Frigid Siberia, Russia. (Picture from: www.fairfieldnodal.com)

Fig. 14: ZLand being used in the snow, Frigid Siberia, Russia. (Picture from: www.fairfieldnodal.com

Fig. 15: ZLand being used in the snow, Frigid Siberia, Russia. (Picture from: www.fairfieldnodal.com)

Russkoye Oil Field, Russia.

Another demo survey was carried out at the Russkoye oilfield in the western Siberia. The data acquisition there was performed by a division of Yamalgeophysical- Vostok, OAO- "integra Geophysica"

The heavy oil Russkoye field reportedly is one of the largest fields in Russia.

The lowest daily air Temperature recorded during the eight day Russkoye demo survey was between -1.2^0C and $-22.5\ ^0C$.

For the survey, 206 nodes were deployed adjacent to cable Geophone strings, and each node was located next to the station markers. The 2D land line was 5125 meters long, crossing roads, a river and frozen lake in a flat tundra environment.

Over the period of the five days shot, ZLand recorded 3,083 shot points. During deployment, the operators ensured the nodes are forcefully stamp into the snow.

It was also recorded that: all nodes recorded seismic data during the five day period with no mechanical or electronic failures.

The data from nodes were transported into the recorder in SEG D formats and was later formatted in SEG Y format although the company that acquired the data could not process the data because the project is proprietary.

Success about the Project

One hundred percent (100%) of all seismic data were recovered from each node. The survey locale was next to a strong noise generator, the data quality was improved by noise attenuation processing. However, the G. P. S. inside the equipment functioned with no problem despite the snow, tress and other environmental noise.

From evidence of the cited case studies, it is enough to convince the industry of the reliability and high productivity of nodal seismic data acquisition, even in extreme weather conditions. The capability of the nodes to record data continuously for many successive days without any monitoring is highly attractive in this and other environments.

CONCLUSION

Cable-based seismic data acquisition systems have been used forever, are great technology, and will continue to be used for years. However, the new kid on the block, cellular wireless data acquisition, looks bullish and will no doubt become popular with some seismic crews.

Cable-free seismic data acquisition alternatives deserve serious consideration as productivity and HSE demands increase. It is time that we, as an industry, free ourselves from the shackles that cables have burdened us with, and move forward into a new age of truly cable-free seismic data acquisition.

References/Bibliography

Beims T. [Nov. 2009], Technology yields better seismic data, the American Oil and Gas Reporter, (Nov. 2009).

David B. [2011], Cable Vs Cableless: Nodal seismic passes frigid Siberia tests, American Association of Petroleum Geologists (AAPG) Explorer, Sep. 2011, pg. 14, 16, 20.

Dennis F. [July, 2008], Cable- free nodes: the next generation land seismic system, The Leading Edge, July, 2008, pg. 878-881.

Dennis F. [2011], Cable- free Land seismic data acquisition, Oil and Gas Technology, pg. 28- 29.

Gary J. [Jan. 2011], Wireless Seismic Acquisition: Real time Data matters, Wireless Seismics Jan. 2011.

Harry J. [2002], The Future of the Oil and Gas Industry: Past Approaches, new Challenges, World Energy, Vol. 5, No. 3 pg 109- 104

Heath B. [June, 2010], Weighing the role of Cableless and Cable- based systems in the future of land seismic acquisition, First Break, Vol 28, Pg. 69-73.

Louise S. [Oct. 2010], Heading in a new Direction, Oilfield Technology, (Oct. 2010).

Michael H. and Robert S. [2008], Accelerometer Vs. Geophone Response: A Field Case History, 2008 CSPG CSEG CWLS Convention, 2008.

Stefano S. and Umberto S. [July, 2008], Wireless Geophone Networks (WGN) for high density Land acquisition, Dipartimento di Electtronica e Informazione, Politecnico di Milano, July, 2008.

www.fairfieldodal.com

www.wirelessseismics.com